S0-ANR-617

AMAZING INVENTIONS

TELEGRAPH

MARY ELIZABETH SALZMANN

Consulting Editor, Diane Craig, M.A./Reading Specialist

Sandcastle

An Imprint of Abdo Publishing
abdopublishing.com

abdopublishing.com

Published by Abdo Publishing, a division of ABDO, PO Box 398166, Minneapolis, Minnesota 55439.

Printed in the United States of America, North Mankato, Minnesota

062015
092015

Editor: Alex Kuskowski
Content Developer: Nancy Tuminelly
Cover and Interior Design and Production: Mighty Media, Inc.
Photo Credits: Library of Congress, Shutterstock, United States Navy, Wikimedia

Library of Congress Cataloging-in-Publication Data

Salzmann, Mary Elizabeth, 1968- author.
Telegraph / Mary Elizabeth Salzmann ; consulting editor, Diane Craig, M.A./Reading Specialist.
pages cm. -- (Amazing inventions)
Audience: Grades PreK-3.
ISBN 978-1-62403-711-5
1. Telegraph--Juvenile literature. 2. Inventions--History--Juvenile literature. I. Title.
TK5265.S25 2016
621.383--dc23
2014045328

SandCastle™ Level: Fluent

SandCastle™ books are created by a team of professional educators, reading specialists, and content developers around five essential components—phonemic awareness, phonics, vocabulary, text comprehension, and fluency—to assist young readers as they develop reading skills and strategies and increase their general knowledge. All books are written, reviewed, and leveled for guided reading, early reading intervention, and Accelerated Reader™ programs for use in shared, guided, and independent reading and writing activities to support a balanced approach to literacy instruction. The SandCastle™ series has four levels that correspond to early literacy development. The levels are provided to help teachers and parents select appropriate books for young readers.

EMERGING · BEGINNING · TRANSITIONAL · FLUENT

CONTENTS

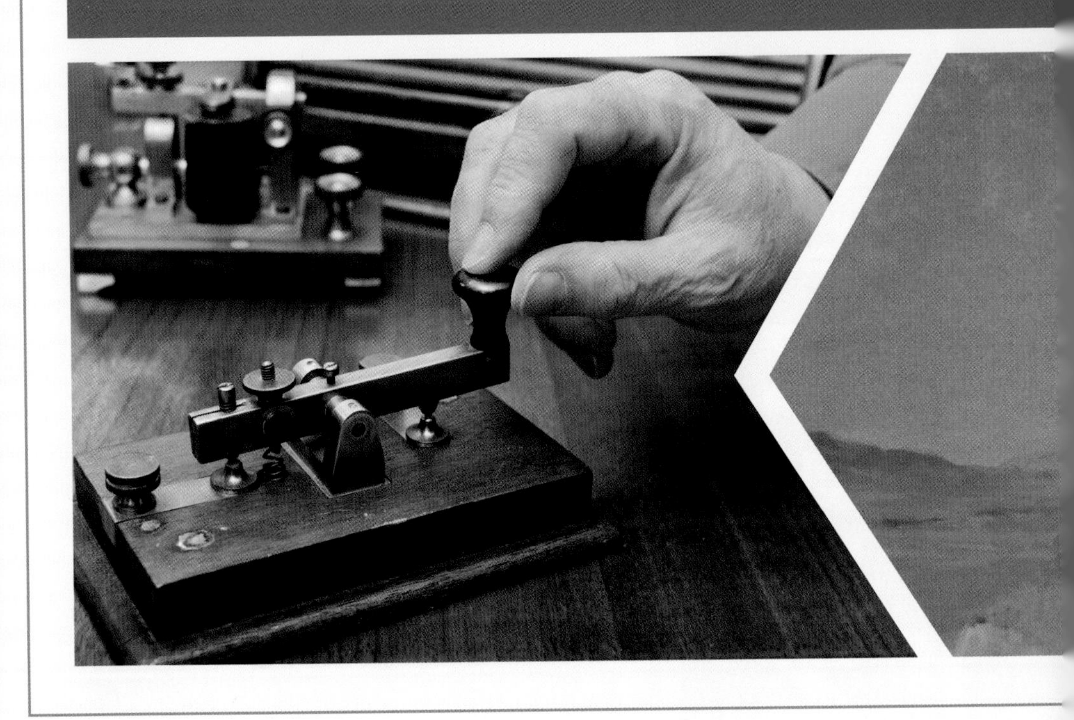

ALL ABOUT TELEGRAPHS

A telegraph is a way to send messages. It uses codes.

Smoke **signals** are a kind of telegraph. People can see them from far away.

Semaphore is also a kind of telegraph.

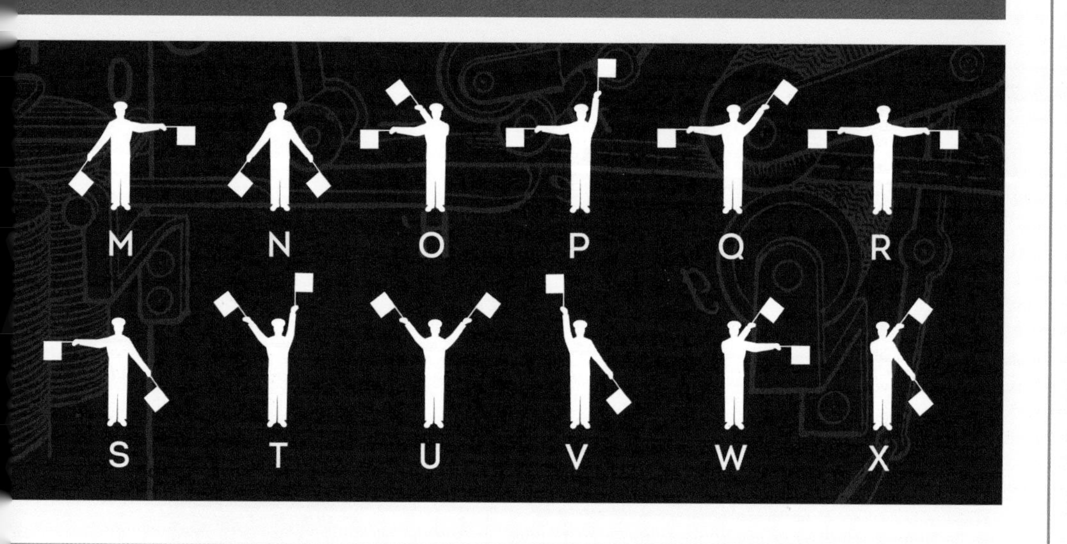

It uses two flags. Each position is a letter.

The US Navy uses semaphore flags.

INTREPID

MORSE CODE

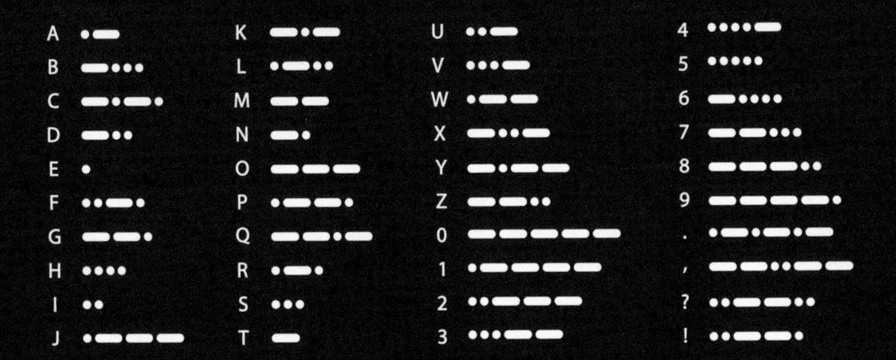

Dots and **dashes** stand for letters.

Samuel Morse built a telegraph machine. He built it in 1837. It was one of the first ones.

Samuel Morse

The sender presses a key.

A short tap is a dot.
A long tap is a **dash**.

The taps travel on wires.
They can go a long way.

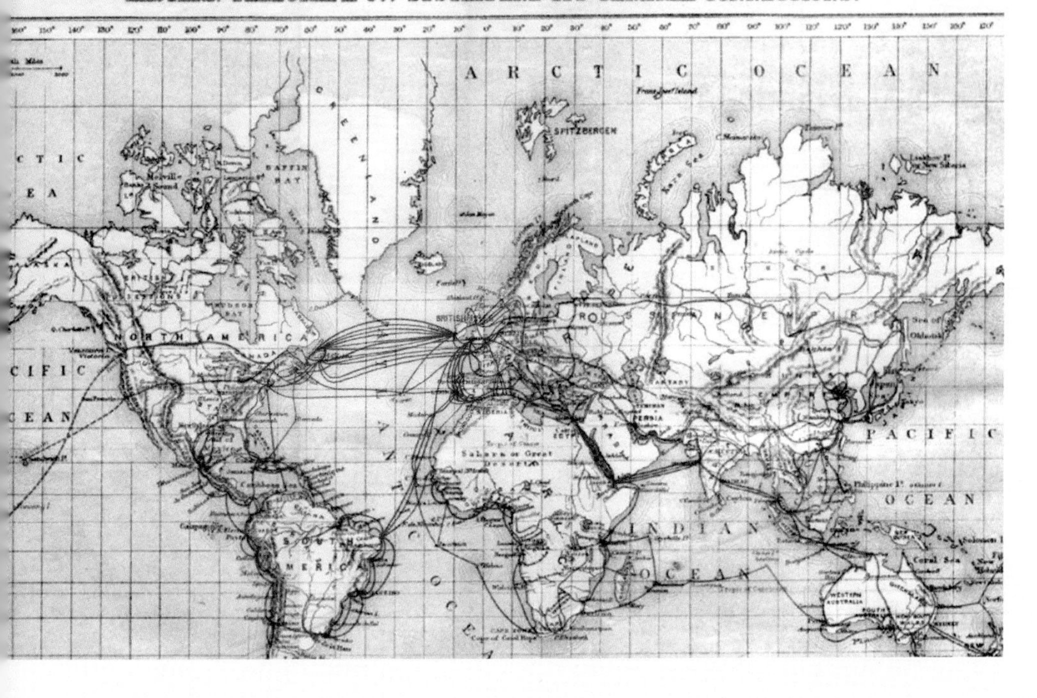

They can even go across the ocean!

The receiver hears the taps.
He or she writes down the letters.

The message is called a **telegram**.

POSTAL TELEGRAPH — COMMERCIAL CABLES

POSTAL TELEGRAPH COMMERCIAL CABLES

TELEGRAM

The Postal Telegraph-Cable Company (Incorporated) transmits and delivers this message subject to the terms and conditions printed on the back of this blank.

COUNTER NUMBER.	TIME FILED. M.	CHECK.

Send the following message, without repeating, subject to the terms and conditions printed on the back hereof, which are hereby agreed to.

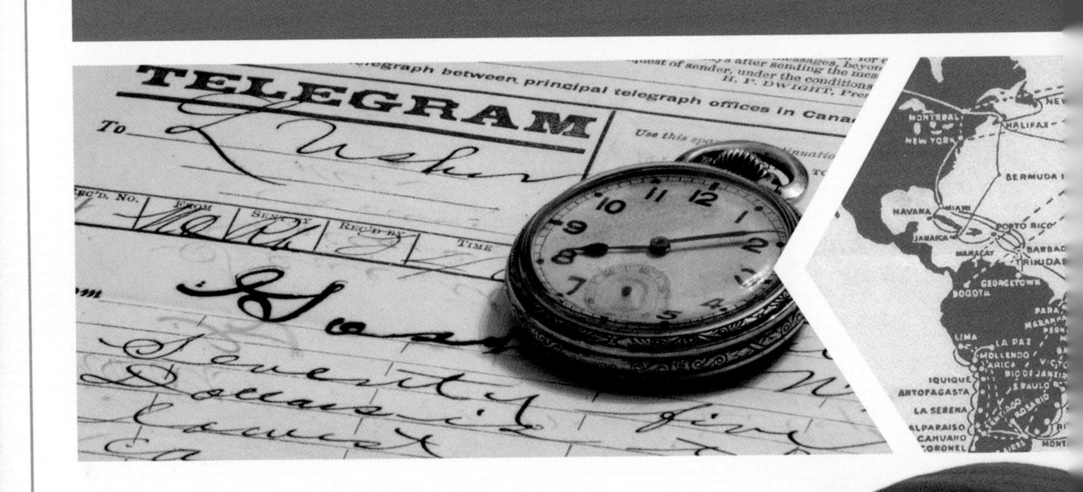

Sending a **telegram** is fast. It is much faster than mailing a letter.

EMDEN
ANTWERP
PARIS
MARSEILLES
LISBON
GIBRALTAR
FAYAL
MADEIRA
LAS PALMAS
TANGIER
TRIPOLI
ALEXANDRIA
CAIRO
TEHERAN
BUSHIRE
KARACHI
CALCUTTA
PEIPING
TOKIO
SHANGHAI
FOO-CHOW
HONGKONG
MANILA
BANGKOK
MADRAS
BOMBAY
COLOMBO
PENANG
LABUAN
PORT SUDAN
ADEN
PERIM
St VINCENT
BATHURST
ACCRA
LAGOS
PRINCIPE
SAN THOME Id
ASCENSION
NAIROBI
MOMBASA
ZANZIBAR
SEYCHELLES Id
BATAVIA
BANJOEWANGI
COCOS Id
BENGUELA
MOSSAMEDES
St HELENA
SALISBURY
MADAGASCAR
MAURITIUS Id
LOURENCO MARQUES
DURBAN
CAPE TOWN
PORT DARWIN
BRISBANE
SYDNEY
PERTH

Telegraphs are not used often today. Now we have phones and **e-mail**.

Send & Receive
receive

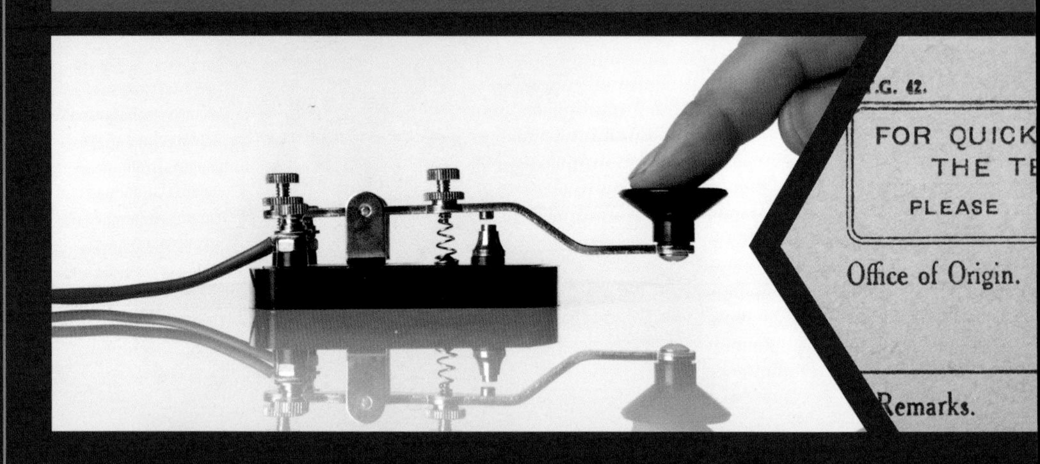

THINK ABOUT IT

Who would you send a **telegram** to? Write your message in Morse code.

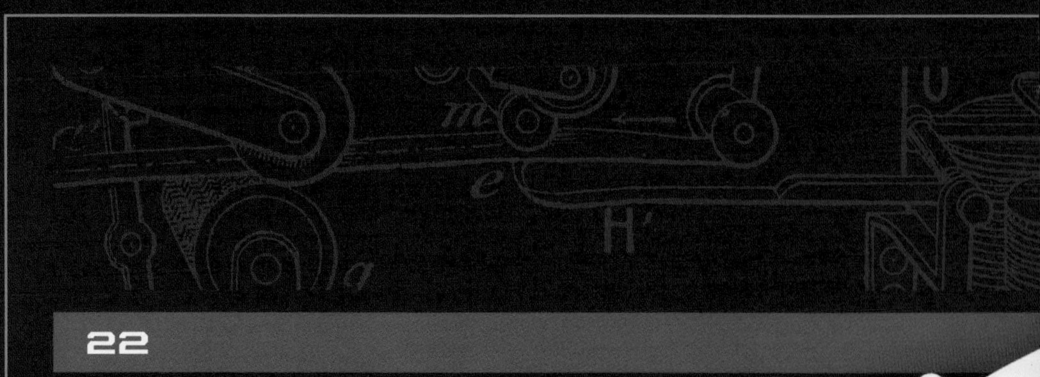

ICE USE
APH.
OVER.

COMMONWEALTH OF AUSTRALIA.—POSTMASTER-GENERAL'S DEPARTMENT.

RECEIVED TELEGRAM

The first line of this telegram contains the following particulars in the order named.

Words. Time Lodged. No.

Ch'nl No

By..........

Sch. C 917.—2/1933.

To

GLOSSARY

dash – a short, horizontal line used in punctuation or Morse code.

e-mail – the system of sending messages using computers, tablets, and smartphones.

signal – a sound, sign, or device that sends a message.

telegram – a message sent through an electric telegraph.